Maritime Drones: Applications and Safety Considerations

Table of Contents

Preface

Introduction

Prelude

Chapter 1: What Are Maritime Drones?

Chapter 2: Key Applications of Maritime Drones

Chapter 3: Military and Commercial Applications

Chapter 4: Technological Advancements and Innovations in Maritime Drones

Chapter 5: Safety Considerations for Maritime Drones

Chapter 6: Legal and Ethical Considerations

Chapter 7: The Future of Maritime Drones

Conclusion

Glossary

Preface

The maritime industry is undergoing a technological revolution, with drones emerging as a transformative tool that has the potential to significantly enhance operations. As technological advancements accelerate, stricter regulations and heightened expectations for operational safety and efficiency drive the industry forward. Recognizing the increasing importance of practical, accessible, and targeted knowledge, the *Gosships Learning Series* was developed to equip maritime professionals with the tools they need to stay ahead of these changes.

This series is designed to provide foundational to intermediate knowledge, with a focus on the real-world application of drone technologies in maritime settings. Coupled with certification tests, each book in the series ensures that the information provided is not only understood but can also be effectively applied in day-to-day operations.

The *Gosships Learning Series* empowers maritime sector personnel—from entry-level crew members to shoreside managers—by equipping them with the skills necessary to navigate the complexities of integrating drones into modern operations. We hope this series will support your professional development and help open new opportunities for growth and success in your career.

Introduction

Welcome to the *Gosships Learning Series*, designed to help maritime professionals expand their knowledge and advance their careers in the rapidly evolving sector. This book, *Maritime Drones: Applications and Safety Considerations*, has been meticulously developed by experienced industry professionals and regulators to ensure that the content is both authoritative and aligned with current industry standards. Whether you are new to the use of drones in maritime operations or seeking to deepen your expertise, this resource is tailored to provide valuable insights into the latest technologies and safety protocols.

In this book, we will explore the following key areas:

- **Applications of Maritime Drones**: Discover how drones are being used in a variety of maritime settings, including inspection of vessels, environmental monitoring, and search and rescue operations.

- **Drone Safety and Regulations**: Understand the regulatory frameworks governing the use of drones in maritime operations and the safety protocols necessary to prevent accidents.

- **Operational Considerations**: Learn how to integrate drones into maritime workflows, from managing drone fleets to ensuring compliance with international and national regulations.

- **Environmental Monitoring**: Explore how drones are revolutionizing the way maritime industries monitor and protect the marine environment.

- **Technological Advancements**: Dive into the latest technological innovations in drone hardware and software, including AI integration and autonomous operation capabilities.

After completing this book, you will be well-prepared to take an assessment designed to test your understanding of maritime drone applications and safety protocols. Upon passing the assessment, you can obtain a Certificate of Achievement by visiting www.gosships.com and accessing the training platform. This certification will validate your knowledge and help demonstrate your expertise to industry peers and employers.

Who Is This Book For?

This book is designed for a wide range of professionals within the maritime industry, including:

- **Maritime and Offshore Personnel**: Individuals looking to expand their technical knowledge on the use of drones in vessel inspections, maintenance, and operations.

- **Shoreside Managers**: Managers overseeing the integration of drones into maritime operations, ensuring regulatory compliance and safety.

- **Aspiring Students**: Those seeking to enter the maritime industry with a solid foundation in modern drone applications and safety considerations.

- **Government and Regulatory Officials**: Personnel responsible for enforcing regulations and ensuring the safe operation of drones in maritime and offshore environments.

By mastering the concepts presented in this book, you will be better equipped to embrace the challenges and opportunities that come with the introduction of drones into maritime operations, stay compliant with international regulations, and contribute to a safer, more efficient industry.

Thank you for choosing the *Gosships Learning Series* to support your journey of continuous learning and professional growth in this exciting and evolving field.

Gosships Learning Series 2024/2025

1. Hydrogen: The Fuel of the Future
2. Green Ammonia: The Next Big Thing in Shipping
3. Decarbonizing Shipping: Pathways to Zero Emissions
4. Battery Technology for Industrial Applications
5. Carbon Capture and Storage: Can It Save the Planet?
6. Biofuels 101: Turning Waste into Energy
7. Understanding LNG (Liquefied Natural Gas)
8. Methanol as a Marine Fuel
9. Offshore Wind Energy: The Future of Renewable Power
10. Tidal and Wave Energy: Harnessing the Ocean
11. Electrofuels: The Next Generation of Carbon-Neutral Fuels
12. Energy Storage Systems for Grid Reliability
13. Hydrogen Fuel Cells for Transportation
14. Solar Energy Innovations: Beyond Solar Panels
15. Smart Grids: The Backbone of Future Energy Systems
16. Ammonia-Hydrogen Blends: A Dual Fuel Solution?
17. Nuclear Power: Small Modular Reactors for a Low-Carbon Future
18. Hydropower: The Oldest Renewable Energy Source
19. Decentralized Energy Systems: Microgrids for Resilience
20. Energy Efficiency Technologies for Industry
21. Hydrogen Production from Seawater
22. Fuel Cells for Maritime Applications
23. Geothermal Energy: Unlocking Earth's Heat
24. Future of EV Charging Infrastructure
25. Synthetic Fuels: Bridging the Gap to Decarbonization
26. Cybersecurity for Maritime and Offshore Operations
27. AI and Automation in Shipping and Logistics
28. Digital Twins in Maritime: Revolutionizing Asset Management

29	Risk Management in Offshore and Maritime Operations
30	Compliance with IMO 2020 Regulations
31	Sustainable Ship Design: Reducing Environmental Impact
32	Marine Renewable Energy: Wave, Tidal, and Offshore Wind Integration
33	Ballast Water Management Systems
34	Blockchain Technology in Shipping: Improving Transpc'y & Efficiency
35	Effective Supply Chain Management for Energy Industries
36	Leadership in the Energy Transition
37	Effective Crisis Management in Maritime Operations
38	Shipyard Safety Management Systems
39	Port State Control (PSC) Inspection Readiness
40	Remote Vessel Operations and Autonomous Shipping
41	Optimizing Fleet Performance with Data Analytics
42	Maritime Environmental Regulations: Staying Ahead of Compliance
43	Advanced Maintenance Strategies: Condition Monitoring & Predictive Maintenance
44	Global LNG Market: Trends and Opportunities
45	Incident Investigation in Maritime Operations
46	International Maritime Law: Key Concepts and Applications
47	Emergency Preparedness and Response for Offshore Oil & Gas
48	Energy Transition Strategies for Oil and Gas Companies
49	Maritime Drones: Applications and Safety Considerations
50	Effective Project Management in Offshore Energy Projects

All Rights Reserved Disclaimer

The contents of this book, including but not limited to all text, graphics, images, logos, and designs, are the intellectual property of Gosships LLC and are protected by copyright law. No part of this publication may be reproduced, distributed, transmitted, displayed, or modified in any form or by any means, including photocopying, recording, or other electronic or mechanical methods, without the prior written permission of the publisher, except in the case of brief quotations in critical reviews or articles.

The information contained within this book is for educational purposes only and is provided "as is" without warranty of any kind, either expressed or implied. The authors and publishers disclaim any liability for any direct, indirect, or consequential loss or damage arising from the use of the material in this book.

For permissions or inquiries, please contact: admin@gosships.com

© 2024 Gosships LLC. All rights reserved.

Prelude

The rapid technological advancements of the 21st century have given rise to unmanned systems across a wide variety of industries. One such development is the deployment of maritime drones in the offshore and shipping sectors. These drones, which operate on the surface, in the air above the sea, and underwater, have revolutionized maritime operations by enhancing efficiency, reducing risk, and enabling innovative approaches to long-standing challenges.

Maritime drones, also referred to as **Unmanned Surface Vehicles (USVs)**, **Unmanned Aerial Vehicles (UAVs)**, and **Unmanned Underwater Vehicles (UUVs)**, are used for applications ranging from environmental monitoring to search and rescue missions. While the technology holds significant potential for improving maritime operations, the use of drones also raises important questions about safety, legal regulations, and ethical concerns.

This mini-book explores the various applications of maritime drones, focusing on their role in different sectors and discussing critical safety and regulatory considerations. By the end of this book, readers will have a better understanding of how maritime drones are transforming the maritime industry and the safeguards needed to ensure their safe and responsible use.

Chapter 1
What Are Maritime Drones?

1.1 Definition and Types of Maritime Drones

Maritime drones are unmanned vehicles designed to operate in marine environments, either on the surface, underwater, or above the water. They are equipped with sophisticated sensors, cameras, and communication systems to perform a wide range of tasks autonomously or with remote control.

1.1.1 Surface Drones (USVs)

Unmanned Surface Vehicles (USVs) are drones that travel on the surface of the water. These drones are often used for **patrol, navigation**, and **surveillance** missions. They are also employed for **hydrographic surveys**, where they map the seabed using sonar technology.

1.1.2 Underwater Drones (UUVs)

Unmanned Underwater Vehicles (UUVs) operate beneath the ocean surface. These drones are vital for **underwater inspections, scientific research**, and **exploration** in regions inaccessible to humans. They can reach great depths and are used for inspecting offshore oil rigs, monitoring pipelines, and exploring shipwrecks.

1.1.3 Aerial Maritime Drones

Aerial drones, often referred to as Unmanned Aerial Vehicles (UAVs), operate above the water and are typically used for **aerial inspections, cargo transport**, and **search and rescue operations**. UAVs are equipped with high-resolution cameras and sensors that allow them to perform detailed surveillance over large areas, providing real-time data to operators onshore.

1.2 Basic Components of Maritime Drones

Maritime drones are equipped with several key components that enable them to perform complex tasks:

- **Navigation Systems**: These include GPS for surface drones and sonar or inertial navigation systems for underwater drones.
- **Propulsion Systems**: Drones use electric or solar-powered motors to move through the water or air.

- **Sensors and Cameras**: For environmental monitoring, navigation, and data collection.

- **Communication Systems**: Satellite, Wi-Fi, or radio-based communication systems enable drones to send data back to a control center or to operate autonomously.

Chapter 2
Key Applications of Maritime Drones

2.1 Surveillance and Monitoring

One of the primary uses of maritime drones is in the field of **surveillance** and **monitoring**. Maritime drones are deployed to monitor **coastal areas, shipping lanes,** and **offshore installations**. In particular, they play a crucial role in ensuring **port security** and **coastal defense**. These drones can patrol vast areas and detect unusual activities such as unauthorized vessels entering restricted zones or illegal fishing operations.

2.1.1 Maritime Security

Coastal states use drones to enhance maritime security. They provide real-time data on suspicious activities and can be rapidly deployed to respond to incidents such as smuggling, piracy, and illegal fishing. Drones equipped with night vision and thermal imaging cameras offer enhanced capabilities, especially during nighttime operations.

2.2 Search and Rescue Operations

Drones have become indispensable tools in **search and rescue missions**. They can cover large areas quickly, often reaching remote locations faster than manned vessels or aircraft. Aerial drones equipped with **infrared sensors** can locate survivors in the water, while surface drones can deploy flotation devices to assist victims. Their ability to operate in adverse conditions, such as rough seas or low visibility, makes them invaluable in emergency situations.

2.2.1 Enhancing Response Times

In many cases, drones reduce the time required to locate and assist distressed vessels or individuals. By sending real-time data and imagery back to rescue coordination centers, drones improve the decision-making process and allow authorities to deploy rescue resources more efficiently.

2.3 Environmental Monitoring

Maritime drones are extensively used for **environmental monitoring**. UUVs are particularly effective for collecting data on water quality, pollution levels, and ocean currents. USVs and UAVs are used to monitor **marine wildlife**, track **climate change indicators**, and assess the impact of **human activities** on marine ecosystems.

2.3.1 Pollution Detection and Control

Maritime drones play an essential role in detecting and monitoring **oil spills**, **chemical leaks**, and **waste discharge** in oceans and coastal areas. By providing real-time data, drones allow authorities to quickly respond to environmental threats and mitigate the impact of pollution.

2.4 Infrastructure Inspection and Maintenance

Drones are also employed for inspecting offshore structures such as **oil rigs**, **wind farms**, and **pipelines**. Traditionally, these inspections required human divers or expensive equipment. Maritime drones, however, can perform these tasks more safely and at a lower cost. UUVs can navigate through underwater structures to detect corrosion or damage, while UAVs provide aerial inspections of offshore wind turbines and oil platforms.

2.4.1 Reducing Human Risk

Using drones for inspections reduces the need for human divers, who often face hazardous conditions when inspecting underwater structures. Drones can access tight or dangerous areas with greater ease and precision, making inspections faster and safer.

Chapter 3
Military and Commercial Applications

3.1 Military Use of Maritime Drones

Maritime drones are transforming military operations at sea. The **naval forces** of many countries deploy drones for **intelligence gathering, surveillance, reconnaissance**, and **mine detection**. Unmanned systems reduce the risk to human personnel in conflict zones, particularly when navigating contested waters or hostile environments.

3.1.1 Mine Detection and Neutralization

Underwater drones are invaluable for detecting underwater mines in shipping lanes and harbors. Equipped with sonar and advanced detection technologies, they can safely identify mines without risking the lives of human divers.

3.1.2 Underwater Warfare

In addition to surveillance, UUVs are used for **underwater warfare**. These drones can engage in covert operations, track enemy submarines, or disrupt communication cables, providing an advantage to naval forces in both offensive and defensive operations.

3.2 Commercial Shipping and Logistics

In the commercial shipping sector, maritime drones are used to improve **cargo transport**, **survey shipping routes**, and manage **port operations**. Aerial drones provide real-time monitoring of vessel traffic in busy ports, while surface drones can autonomously navigate shipping lanes to deliver goods between ports.

3.2.1 Autonomous Cargo Ships

The future of maritime logistics lies in the development of **autonomous cargo ships**. These vessels will rely on drones for **navigational support**, **inspection**, and **communication**. Maritime drones will serve as "eyes and ears" for autonomous ships, ensuring safe navigation and efficient cargo handling.

Chapter 4

Technological Advancements and Innovations in Maritime Drones

4.1 Navigation and Autonomy

Advances in **GPS**, **AI**, and **sensor technology** have significantly improved the autonomy of maritime drones. Autonomous drones can now navigate complex environments, avoid obstacles, and make real-time decisions without human intervention. The integration of **machine learning algorithms** enables drones to adapt to changing conditions and optimize their navigation routes.

4.1.1 AI and Machine Learning

Artificial intelligence (AI) plays a crucial role in enhancing the decision-making capabilities of drones. AI allows drones to process large amounts of data, identify patterns, and make adjustments in real time. For instance, a drone used for environmental monitoring can adjust its course based on changing weather conditions or ocean currents.

4.2 Communication Systems

Effective communication is critical for the successful operation of maritime drones, especially in remote areas. Drones rely on a variety of communication systems, including **satellite links**, **radio waves**, and **Wi-Fi**. However, maintaining stable communication is a challenge in extreme environments such as deep waters or distant oceanic regions.

4.2.1 Satellite Communication

Satellites provide reliable communication links for drones operating far from shore. This allows real-time data transmission between drones and control centers, enabling operators to monitor operations and make critical decisions.

4.3 Power Systems

The range and endurance of maritime drones depend heavily on their power systems. Advances in **battery technology, solar-powered drones**, and **energy-efficient propulsion systems** have extended the operational capabilities of drones, allowing them to perform longer missions with minimal downtime.

4.3.1 Renewable Energy Integration

The integration of renewable energy sources, such as **solar panels** on drone surfaces, has enabled continuous power generation during long-distance maritime operations. This allows drones to remain operational for extended periods, particularly in areas with ample sunlight.

Chapter 5
Safety Considerations for Maritime Drones

5.1 Collision Avoidance and Navigation Safety

Ensuring that drones avoid collisions with other vessels, structures, or marine wildlife is a top priority. Modern drones are equipped with **collision avoidance systems**, which use sensors and real-time data to detect potential hazards and adjust the drone's path accordingly.

5.1.1 Sensor-Based Avoidance

Drones use a combination of radar, sonar, and cameras to detect obstacles. These sensors allow drones to calculate safe distances from other objects and adjust their course to avoid collisions. In crowded waters, such as ports or shipping lanes, collision avoidance is essential for preventing accidents.

5.2 Regulatory Compliance

The use of drones in maritime operations is subject to various national and international regulations. These regulations, developed by organizations like the **International Maritime Organization (IMO)** and **regional maritime authorities**, set guidelines for the safe deployment of drones in commercial and military operations.

5.2.1 Aligning with Maritime Law

One challenge is that drone technology has outpaced regulatory frameworks in many regions. Operators must ensure that drone usage complies with existing maritime laws, including regulations governing airspace, shipping lanes, and port security. Regulatory bodies are working to update guidelines to accommodate the growing role of drones in maritime industries.

5.3 Environmental and Operational Risks

Drones operating in marine environments must account for the potential risks posed by adverse weather conditions, such as storms, high winds, and rough seas. Additionally, there is growing concern over the **environmental impact** of drones on marine ecosystems.

5.3.1 Minimizing Disruption to Marine Wildlife

The noise generated by drones, particularly UUVs, can disturb marine animals such as whales and dolphins. Drone operators must take steps to mitigate these impacts by adhering to wildlife protection guidelines and using drones responsibly in sensitive habitats.

5.4 Data Security and Privacy

Maritime drones are often used to collect sensitive data, whether for commercial, military, or environmental purposes. Ensuring that this data remains secure is a critical consideration. **Cybersecurity** measures must be in place to protect drones from hacking or interference.

5.4.1 Protecting Communication Channels

Drones use encrypted communication channels to transmit data. However, these channels are vulnerable to cyberattacks, particularly in regions with geopolitical tensions. Operators must invest in robust security protocols to safeguard drone data and communication.

Chapter 6
Legal and Ethical Considerations

6.1 Liability and Accountability

With the increased use of drones in maritime operations comes the question of **liability**. In the event of a drone accident or incident, it is essential to establish who is responsible—whether it's the operator, the drone manufacturer, or the owner of the vessel or infrastructure involved.

6.1.1 Drone Accidents and Property Damage

If a drone causes damage to a ship or offshore platform, determining liability is complex. In many cases, insurance policies may cover the damages, but establishing accountability can be a lengthy process.

6.2 Ethical Use of Maritime Drones

The ethical use of drones, particularly for **surveillance** and **military applications**, raises important concerns. Drones have the potential to infringe on privacy, especially when used for **monitoring private vessels** or **coastal areas**.

6.2.1 Surveillance and Privacy Issues

Maritime drones used for security purposes must balance the need for surveillance with the protection of individual privacy. Ethical guidelines must be established to ensure that drones are not used for unwarranted or invasive monitoring of private vessels or sensitive areas.

Chapter 7
The Future of Maritime Drones

7.1 Expansion of Autonomous Vessels

The development of fully autonomous ships is expected to be one of the most significant innovations in the maritime industry. Drones will play a crucial role in the operation of these vessels, providing navigation, maintenance, and surveillance support. As drone technology advances, autonomous ships could become a reality within the next decade.

7.1.1 Autonomous Fleets

In the future, entire fleets of autonomous surface vessels and underwater drones may operate in concert, managed by a central control system. These fleets will be capable of performing complex missions without human intervention, such as surveying large ocean regions or transporting goods across the globe.

7.2 Integration with Smart Ports and Smart Shipping

Maritime drones will also be key components of **smart port infrastructure**. In smart ports, drones will automate cargo handling, traffic management, and security. They will work alongside other intelligent systems, such as automated cranes and self-driving trucks, to create highly efficient and sustainable port operations.

7.2.1 Data-Driven Shipping

Smart shipping relies on real-time data collection and analysis. Drones equipped with sensors will provide continuous updates on vessel conditions, weather patterns, and port traffic, optimizing shipping routes and reducing delays.

7.3 Potential for AI-Driven Drone Swarms

One emerging trend in drone technology is the development of **drone swarms**, where multiple drones work together to complete a task. In maritime applications, AI-driven drone swarms could be used for large-scale **environmental monitoring, search and rescue**, or **military operations**.

7.3.1 Collaborative Drones

By communicating with each other in real time, drone swarms can coordinate complex missions, such as monitoring vast areas of the ocean for environmental changes or conducting a simultaneous inspection of multiple offshore structures. The ability of drone swarms to collaborate and share data enhances their efficiency and effectiveness.

Conclusion

The integration of drones into maritime operations offers numerous advantages, from enhanced surveillance and search-and-rescue capabilities to more efficient port operations and infrastructure inspections. However, the rapid advancement of drone technology also brings with it a set of challenges, particularly regarding safety, regulatory compliance, and ethical concerns.

As the maritime industry continues to adopt and develop these technologies, it is essential to strike a balance between innovation and responsibility. By adhering to safety standards, ensuring environmental sustainability, and addressing privacy and liability issues, the full potential of maritime drones can be realized in the years to come.

Glossary: Maritime Drones: Applications and Safety Considerations

1. **AI (Artificial Intelligence)**: Technology enabling drones to process data and make decisions autonomously during maritime operations, improving efficiency and safety.

2. **Autonomous Drones**: Drones that operate without human intervention, relying on pre-programmed instructions or AI to perform tasks like inspections and surveillance.

3. **Battery Life**: The duration a drone can operate on a single charge, critical for planning long-distance or extended maritime missions.

4. **Beyond Visual Line of Sight (BVLOS)**: A regulatory classification allowing drone operators to control drones beyond their direct line of sight, essential for wide-area maritime operations.

5. **Calibration**: The process of adjusting a drone's sensors and cameras to ensure accurate data collection and performance during maritime missions.

6. **Collision Avoidance System**: A technology that helps drones detect and avoid obstacles during flight, critical for safe navigation in congested maritime environments.

7. **Data Transmission**: The process of sending collected data from a drone to a control center or operator, often requiring reliable communication systems, especially in remote maritime areas.

8. **Digital Twin**: A virtual representation of a physical asset, such as a vessel or offshore platform, created using drone-collected data for real-time monitoring and predictive maintenance.

9. **Drone Fleet Management**: The coordination and oversight of multiple drones to optimize their performance in various maritime tasks, such as inspections and environmental monitoring.

10. **Drone Port**: A designated location, often on ships or offshore platforms, where drones can take off, land, and recharge between missions.

11. **Drone Regulations**: The rules and guidelines governing the safe and legal operation of drones in maritime environments, set by international bodies like the IMO or national aviation authorities.

12. **Electro-Optical (EO) Sensor**: A camera sensor used by drones to capture high-resolution imagery during daylight conditions, often used for vessel and port inspections.

13. **Emergency Response Drones**: Drones specifically designed or deployed for emergency situations such as search and rescue, firefighting, or oil spill detection.

14. **Environmental Impact Assessment (EIA)**: A process where drones are used to monitor and assess the environmental effects of maritime operations, such as pollution or habitat disruption.

15. **Failsafe System**: A built-in safety mechanism in drones that automatically returns the drone to a safe location or lands it safely in case of system failure or signal loss.

16. **Flight Altitude**: The height at which a drone operates during maritime missions, regulated to ensure safety and proper data capture.

17. **Flight Path**: The planned route a drone will take during its operation, often pre-programmed or controlled in real-time to optimize mission objectives.

18. **Geofencing**: A safety feature that uses GPS to create virtual boundaries, preventing drones from entering restricted maritime zones or airspaces.

19. **GIS (Geographic Information System)**: A system that captures, stores, and analyzes spatial data, often used by drones for mapping and environmental monitoring in maritime settings.

20. **Hydrography**: The science of measuring and mapping underwater features, where drones equipped with sonar systems can assist in collecting data for safe navigation.

21. **IMO (International Maritime Organization)**: A specialized UN agency that sets global standards for the safety, security, and environmental performance of maritime operations, including drone usage.

22. **Inspection Drone**: A drone equipped with cameras and sensors to inspect ships, offshore platforms, and other maritime infrastructure for damage or regulatory compliance.

23. **IP Rating (Ingress Protection)**: A standard defining a drone's ability to withstand dust and water, crucial for drones operating in harsh maritime environments.

24. **Landing Zone**: A predetermined area where a drone is programmed to land safely, either on a vessel, platform, or shore.

25. **LiDAR (Light Detection and Ranging)**: A remote sensing method that uses light in the form of a pulsed laser to measure distances, often used by drones for 3D mapping in maritime applications.

26. **Line of Sight (LOS)**: The operator's ability to maintain visual contact with the drone during flight, required by many regulations for safe operation.

27. **Maritime Surveillance Drones**: Drones used for monitoring and surveillance in maritime environments, tracking vessel movements, illegal fishing, or environmental changes.

28. **Multirotor Drones**: Drones with multiple rotors, typically used in maritime operations for their stability and ability to hover, ideal for inspections and close-up monitoring.

29. **Navigation System**: The onboard system that allows a drone to determine its position and follow a flight path accurately, crucial for maritime operations in open water.

30. **Payload**: The equipment a drone carries, such as cameras, sensors, or environmental monitors, which varies depending on the maritime mission.

31. **Payload Capacity**: The maximum weight of equipment a drone can carry without affecting its flight performance, important for determining the drone's operational capabilities.

32. **Remote Sensing**: The collection of data from a distance, typically using drone-mounted sensors, to monitor environmental conditions or maritime assets.

33. **Return-to-Home (RTH)**: An automated drone function that directs the drone to return to its starting point or a pre-designated safe area when the mission is complete or in case of an emergency.

34. **RF (Radio Frequency) Communication**: The method used by drones to communicate with their controllers or data centers, vital for maintaining control over long distances in maritime operations.

35. **Risk Assessment**: The process of identifying and mitigating potential risks associated with drone operations in maritime environments, such as weather conditions or navigational hazards.

36. **Safety Management System (SMS)**: A structured approach to managing safety in drone operations, ensuring compliance with regulations and minimizing risks during maritime missions.

37. **Satellite Communication (SatCom)**: A communication system used by drones in remote maritime areas where traditional communication signals are weak or unavailable.

38. **Search and Rescue (SAR)**: The use of drones equipped with thermal imaging or other sensors to locate people in distress during maritime emergencies.

39. **Single-Rotor Drones**: Drones with a single rotor, typically larger and more powerful, used for long-range maritime missions or carrying heavy payloads.

40. **Situational Awareness**: The real-time knowledge of conditions and potential hazards in the drone's operational environment, critical for safe maritime drone operations.

41. **Sonar**: A technology that uses sound waves to detect underwater objects or measure water depth, often used in drones for underwater inspections and hydrography.

42. **Telemetry**: The transmission of real-time data from the drone to the operator, including information on flight status, battery levels, and environmental conditions.

43. **Thermal Imaging Camera**: A camera that detects heat and is often used by drones in search and rescue missions or to identify hotspots on vessels.

44. **Unmanned Aerial Vehicle (UAV)**: A drone that operates without a human pilot on board, used in maritime operations for tasks such as inspections, surveillance, and data collection.

45. **Unmanned Traffic Management (UTM)**: A system designed to manage the movement and operation of multiple drones within a given airspace, especially important in busy maritime areas.

46. **Vessel Traffic Service (VTS)**: A marine monitoring system designed to manage and track vessel traffic in busy waters, where drones can assist in ensuring safety and efficiency.

47. **Visual Line of Sight (VLOS)**: The requirement that a drone operator must keep the drone within their direct line of sight during flight to ensure safe operation.

48. **VTOL (Vertical Takeoff and Landing)**: A drone capable of taking off and landing vertically, making it ideal for use in maritime environments where space is limited.

49. **Weatherproofing**: The process of protecting a drone from adverse weather conditions such as rain, wind, and saltwater corrosion, essential for drones operating in maritime environments.

50. **Waypoint Navigation**: A flight mode in which a drone follows a series of predefined GPS coordinates, commonly used in maritime drone operations for precision tasks such as surveying or inspections.

www.ingramcontent.com/pod-product-compliance
Lightning Source LLC
Chambersburg PA
CBHW030041230526
45472CB00002B/625